Dimensions M
Textbook PKB

Authors and Reviewers
Tricia Salerno
Pearly Yuen
Jenny Kempe
Cassandra Turner
Allison Coates

Consultant
Dr. Richard Askey

Singapore Math Inc.

Published by Singapore Math Inc.

19535 SW 129th Avenue
Tualatin, OR 97062
www.singaporemath.com

Dimensions Math® Textbook Pre-Kindergarten B
ISBN 978-1-947226-01-2

First published 2018
Reprinted 2019, 2020, 2021, 2023 (twice)

Copyright © 2017 by Singapore Math Inc.
All rights reserved. This book or any portion thereof may not be reproduced or used in any manner whatsoever without the express written permission of the publisher.

Printed in China

Acknowledgments

Design and illustration by Cameron Wray.

Preface

The Dimensions Math® Pre-Kindergarten to Grade 5 series is based on the pedagogy and methodology of math education in Singapore. The curriculum develops concepts in increasing levels of abstraction, emphasizing the three pedagogical stages: Concrete, Pictorial, and Abstract. Each topic is introduced, then thoughtfully developed through the use of exploration, play, and opportunities for mastery of skills.

Features and Lesson Components

Students work through the lessons with the help of five friends: Emma, Alex, Sofia, Dion, and Mei. The characters introduce themselves in Pre-K and continue to appear throughout the series. They give instructions, hints, and ideas.

Chapter Opener

Each chapter begins with an engaging scenario that stimulates student curiosity in new concepts. This scenario also provides teachers an opportunity to review skills.

Lesson

Engaging pictures draw the students into the concept of each lesson.

Exercise

A pencil icon ✏️ at the end of the lesson links to additional practice problems in the workbook.

Review

A review of chapter material provides ongoing practice of concepts and skills.

Note: There are additional lesson components in the teacher's guide: Explore, Learn, Play, and Extend.

Contents

Chapter		Lesson	Page
Chapter 8 **Ordinal Numbers**		Chapter Opener	1
	1	First	2
	2	Second and Third	3
	3	Fourth and Fifth	4
	4	Practice	5
Chapter 9 **Shapes and Solids**		Chapter Opener	7
	1	Cubes, Cylinders, and Spheres	8
	2	Cubes	10
	3	Positions	11
	4	Build with Solids	13
	5	Rectangles and Circles	15
	6	Squares	17
	7	Triangles	18
	8	Squares, Circles, Rectangles, and Triangles — Part 1	19
	9	Squares, Circles, Rectangles, and Triangles — Part 2	21
	10	Practice	22

Chapter		Lesson	Page
Chapter 10 **Compare Sets**		Chapter Opener	25
	1	Match Objects	26
	2	Which Set Has More?	30
	3	Which Set Has Fewer?	33
	4	More or Fewer?	36
	5	Practice	39
Chapter 11 **Compose and Decompose**		Chapter Opener	43
	1	Altogether — Part 1	44
	2	Altogether — Part 2	45
	3	Show Me	48
	4	What's the Other Part? — Part 1	51
	5	What's the Other Part? — Part 2	53
	6	Practice	55

Chapter	Lesson	Page

Chapter 12
Explore Addition and Subtraction

	Chapter Opener	59
1	Add to 5 — Part 1	60
2	Add to 5 — Part 2	61
3	Two Parts Make a Whole	63
4	How Many in All?	65
5	Subtract Within 5 — Part 1	66
6	Subtract Within 5 — Part 2	68
7	How Many Are Left?	69
8	Practice	70

Chapter	Lesson	Page
Chapter 13 **Cumulative Review**	Chapter Opener	73
	Review 1 Match and Color	74
	Review 2 Big and Small	77
	Review 3 Heavy and Light	78
	Review 4 Count to 5	79
	Review 5 Count 5 Objects	81
	Review 6 0	82
	Review 7 Count Beads	83
	Review 8 Patterns	85
	Review 9 Length	86
	Review 10 How Many?	87
	Review 11 Ordinal Numbers	89
	Review 12 Solids and Shapes	90
	Review 13 Which Set Has More?	94
	Review 14 Which Set Has Fewer?	95
	Review 15 Put Together	96
	Review 16 Subtraction	98
	Looking Ahead 1 Sequencing — Part 1	99
	Looking Ahead 2 Sequencing — Part 2	100
	Looking Ahead 3 Categorizing	102
	Looking Ahead 4 Addition	103
	Looking Ahead 5 Subtraction	105
	Looking Ahead 6 Getting Ready to Write Numerals	107
	Looking Ahead 7 Reading and Math	108

Chapter 8

Ordinal Numbers

Lesson 1
First

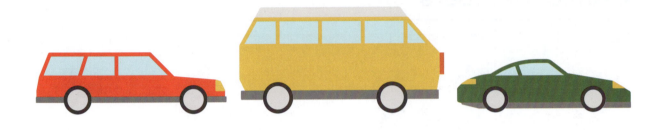

Circle the first in line from the front.

Objective: Identify first from the front.

Exercise 1 • page 1

Lesson 2
Second and Third

What color is the first car from the front?
What color is the second car from the front?
What color is the third car from the front?

Color the first car from the front in each train green.
Color the second car from the front in each train yellow.
Color the third car from the front in each train orange.

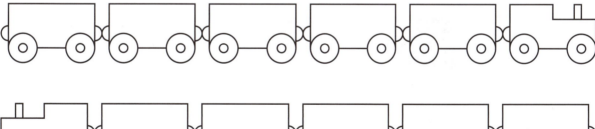

Objective: Identify second and third from the front.

Exercise 2 • page 3

Lesson 3
Fourth and Fifth

Circle the fourth bead from the top.
Cross out the fifth bead from the bottom.

Objective: Identify fourth and fifth from a starting point.

Exercise 3 • page 5

Lesson 4
Practice

Color the first butterfly from the left red.
Color the second butterfly from the left blue.
Color the third butterfly from the left purple.

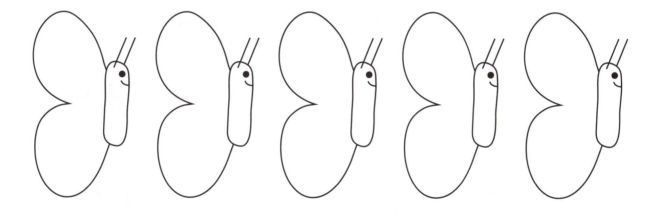

Objective: Practice.

Circle the first bird in each row.
Cross out the fourth bird in each row.
Draw a box around the fifth bird in each row.

Objective: Practice.

Exercise 4 • page 7

Chapter 9

Shapes and Solids

Lesson 1
Cubes, Cylinders, and Spheres

Cross out the one that does not belong in each row.

Objective: Identify some solids.

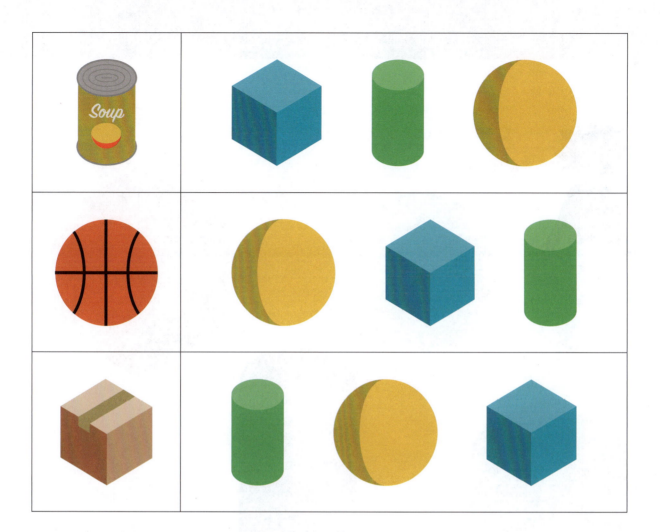

Circle the solid that is shaped like the solid on the left in each row.

Objective: Identify some solids.

Exercise 1 • page 9

9-1 Cubes, Cylinders, and Spheres

Lesson 2
Cubes

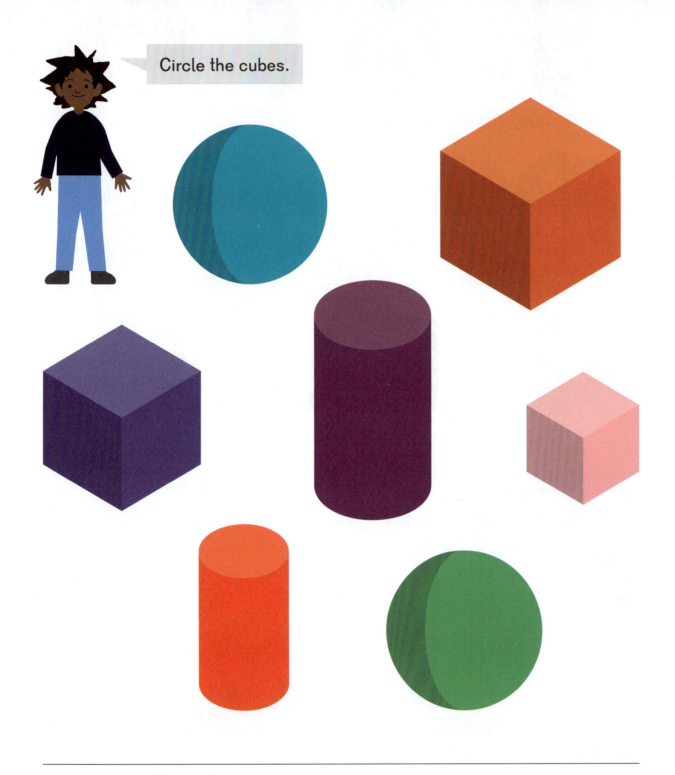

Circle the cubes.

Objective: Identify cubes.

Exercise 2 • page 11

Lesson 3
Positions

My pink cube is at the top of my tower.
My orange cube is at the bottom of the tower.
What color cubes do you see in between the pink and the orange linking cubes?

Objective: Learn positional words.

This is a picture of my bear, Brownie.
Color the block above Brownie's picture red.
Color the block in front of Brownie's picture green.
Color the block behind Brownie's picture black.
Color the block beside Brownie's picture purple.

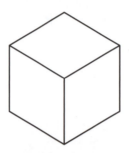

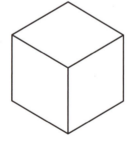

Objective: Learn positional words.

Exercise 3 • page 13

Lesson 4
Build with Solids

Objective: Build using some solids.

Circle the cube that is between two other cubes.
Cross off a cylinder to the left of the cubes.
Draw a face on the sphere on top of a cylinder.

Objective: Build using some solids.

Lesson 5
Rectangles and Circles

Objective: Identify rectangles and circles.

Objective: Identify rectangles and circles.

Exercise 4 • page 15

Lesson 6
Squares

Objective: Identify squares.

Exercise 5 • page 17

Lesson 7
Triangles

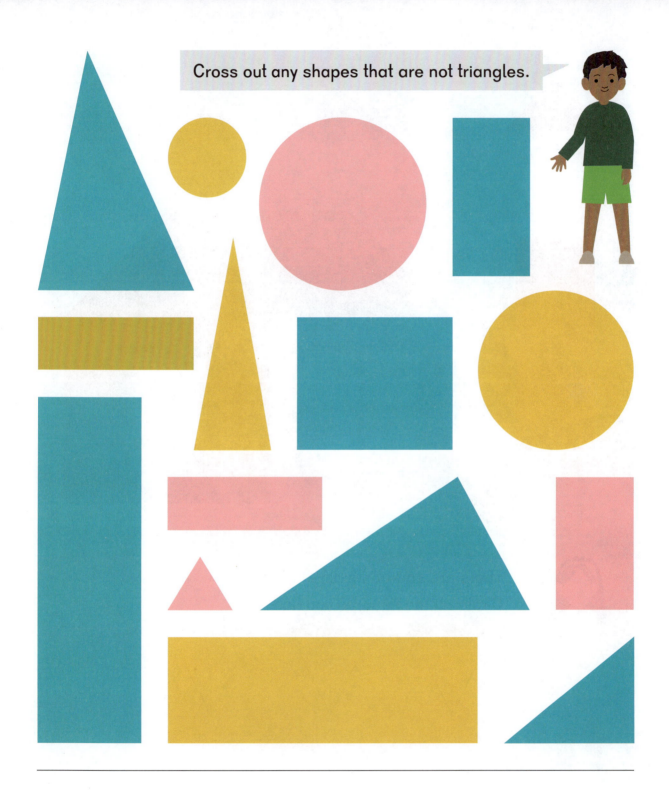

Cross out any shapes that are not triangles.

Objective: Identify triangles.

Exercise 6 • page 19

Lesson 8
Squares, Circles, Rectangles, and Triangles — Part 1

It's a circle and you know it cause it's round.
It's a circle and you know it cause it's round.
It rolls very well.
It's curved you can tell.
It's a circle and you know it cause it's round.
It's a triangle and you know it has 3 sides.
It's a triangle and you know it has 3 sides.
It's also got 3 corners.
Which are pointy, kind of sharp.
It's a triangle and you know it has 3 sides.
It's a rectangle and you know it has 4 sides.
It's a rectangle and you know it has 4 sides.
It's a face on a box.
Now you're clever like a fox.
It's a rectangle and you know it has 4 sides.
A square also has 4 sides.
A square also has 4 sides.
A cube's face is a square.
You can always find it there.
A square also has 4 sides.

Sing with me!

Objective: Learn some attributes of squares, circles, rectangles, and triangles.

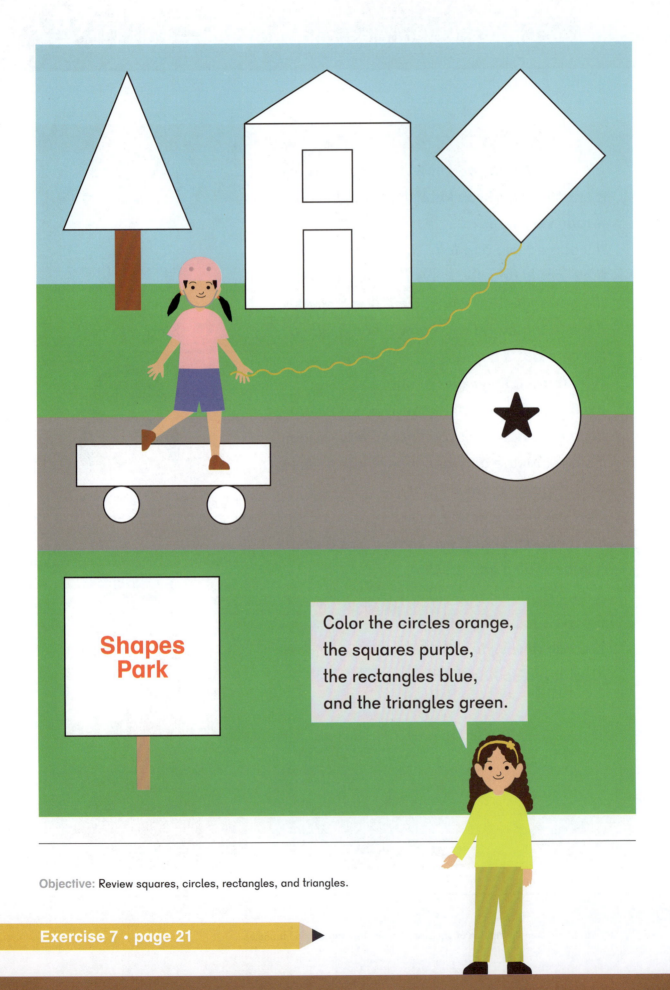

Lesson 9
Squares, Circles, Rectangles, and Triangles — Part 2

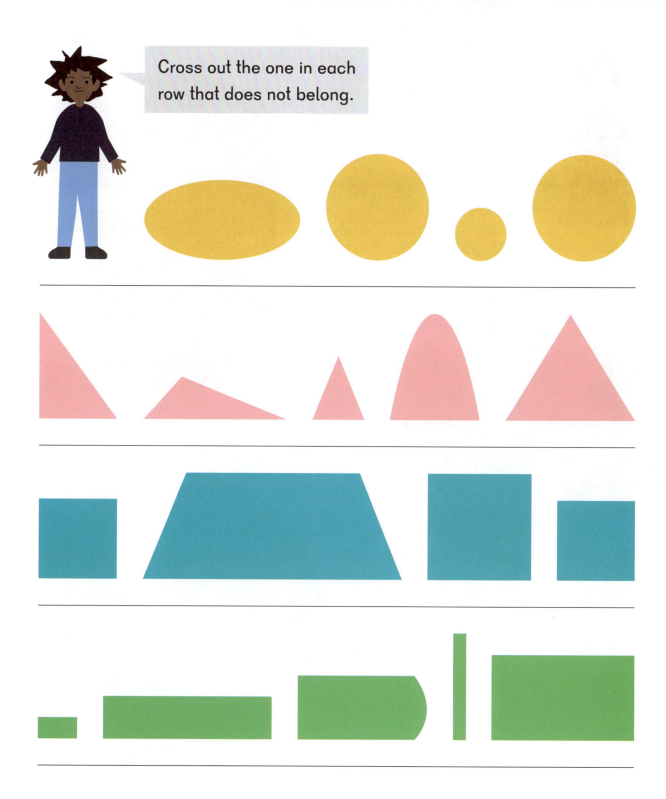

Cross out the one in each row that does not belong.

Objective: Identify examples and counterexamples of squares, circles, rectangles, and triangles.

Exercise 8 • page 23

Lesson 10
Practice

Draw eyes on the cube.
Cross out the sphere.
Circle the cylinder.

Objective: Practice.

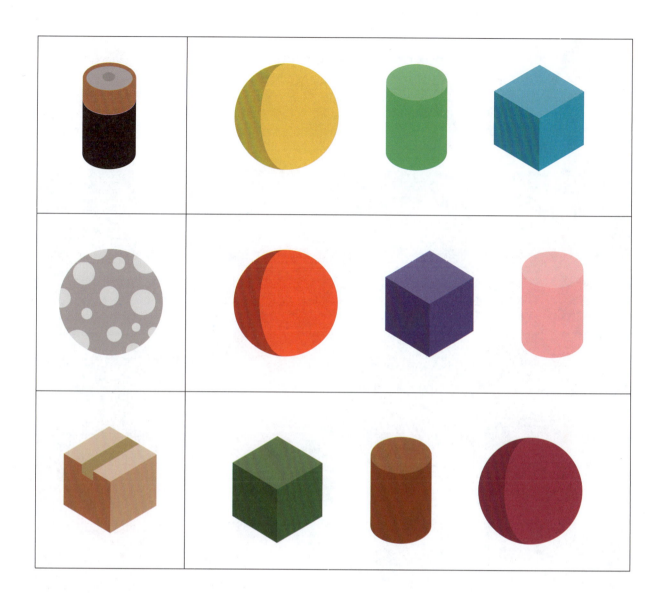

Circle the solid that looks like the object on the left.

Objective: Practice.

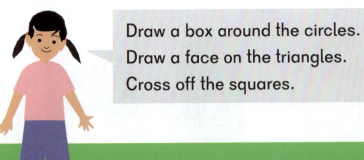

Draw a box around the circles.
Draw a face on the triangles.
Cross off the squares.

Objective: Practice.

Exercise 9 • page 25

Chapter 10

Compare Sets

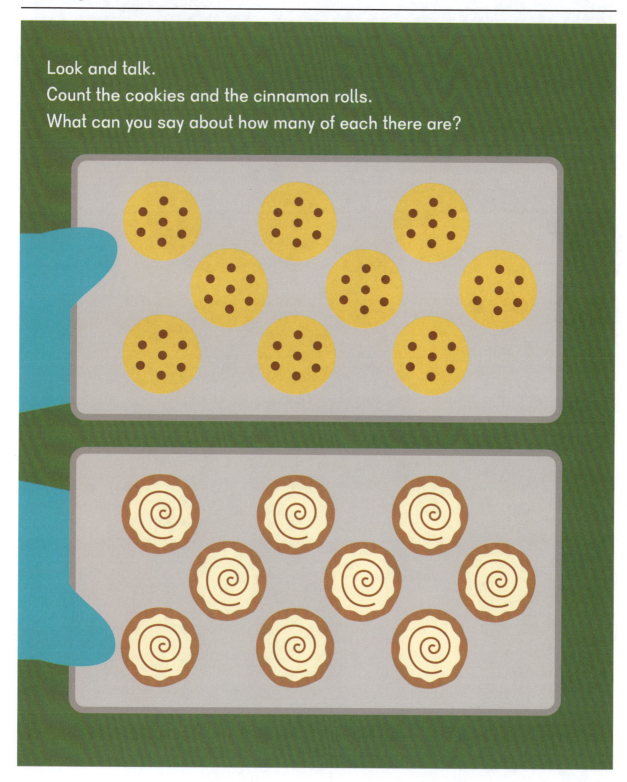

Look and talk.
Count the cookies and the cinnamon rolls.
What can you say about how many of each there are?

Lesson 1
Match Objects

Hey diddle diddle,
The cat and the fiddle,
The cow jumped over the moon.
The little dog laughed
To see such sport,
And the dish ran away with the spoon.

Objective: Match one object to one object.

Match each cat with a fiddle.

Objective: Match one object to one object.

Is there a dish for each spoon?
Use a pencil to match them.

Objective: Match one object to one object.

Objective: Match one object to one object.

Exercise 1 • page 27

Lesson 2
Which Set Has More?

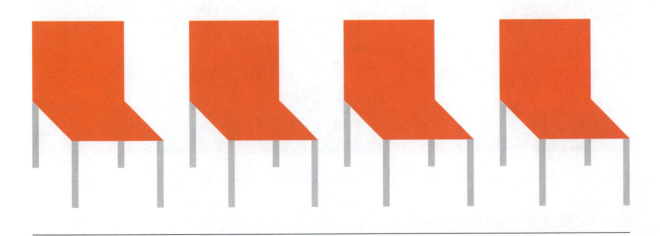

Objective: Compare two sets of objects to find which has more.

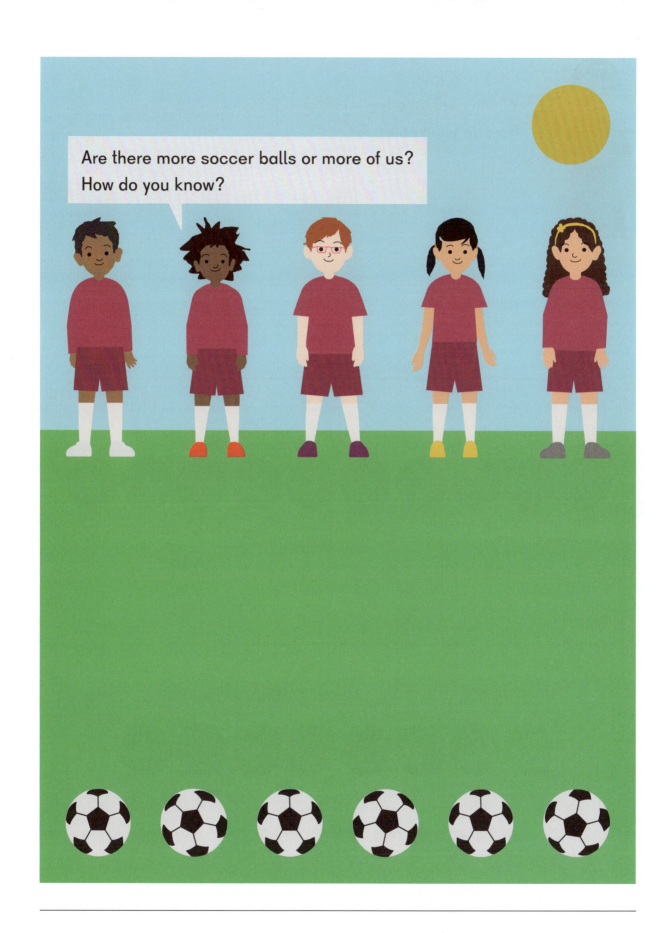

Objective: Compare two sets of objects to find which has more.

Look and talk.
Which necklace has more beads?
How do you know?

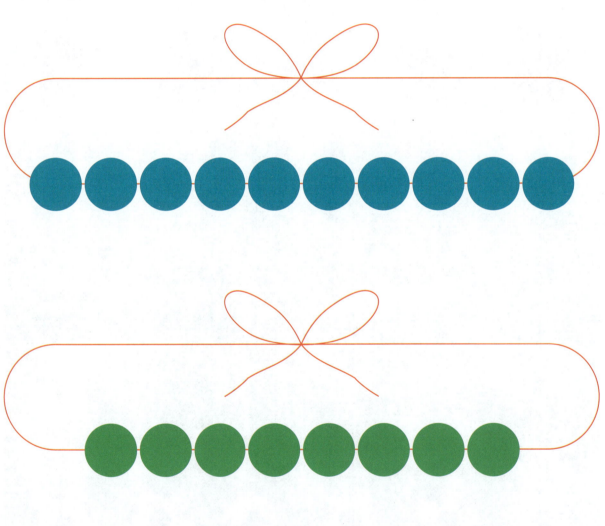

Objective: Compare two sets of objects to find which has more.

Exercise 2 • page 31

Lesson 3
Which Set Has Fewer?

Are there fewer bicycles or tricycles? Circle the group with fewer.

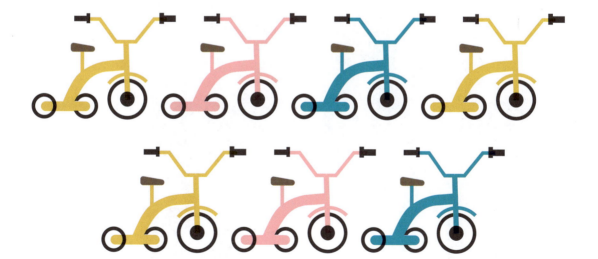

Objective: Compare two sets of objects to find which has fewer.

Which one has fewer flowers? Circle it.

Objective: Compare two sets of objects to find which has fewer.

Exercise 3 • page 33

10-3 Which Set Has Fewer?

Lesson 4
More or Fewer?

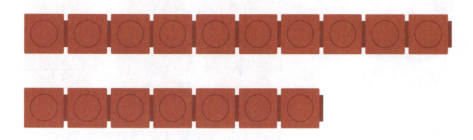

Circle the train that has more cubes.

Objective: Review comparing sets of objects to determine which has more or fewer.

Color the beads on the necklace that has fewer beads.

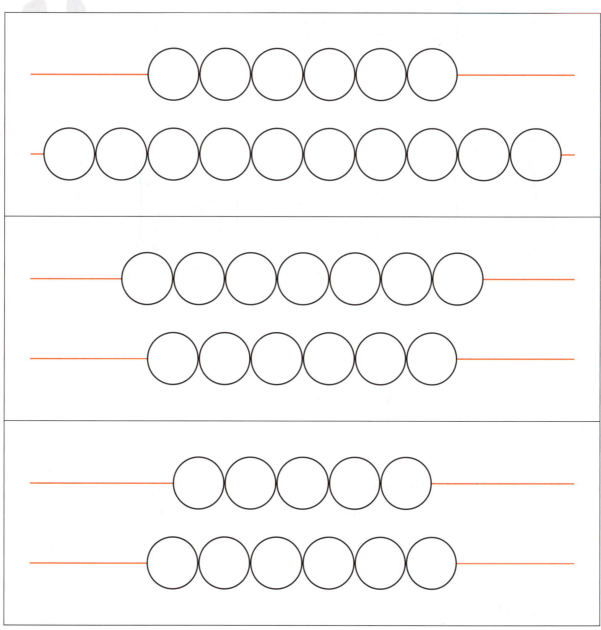

Objective: Review comparing sets of objects to determine which has more or fewer.

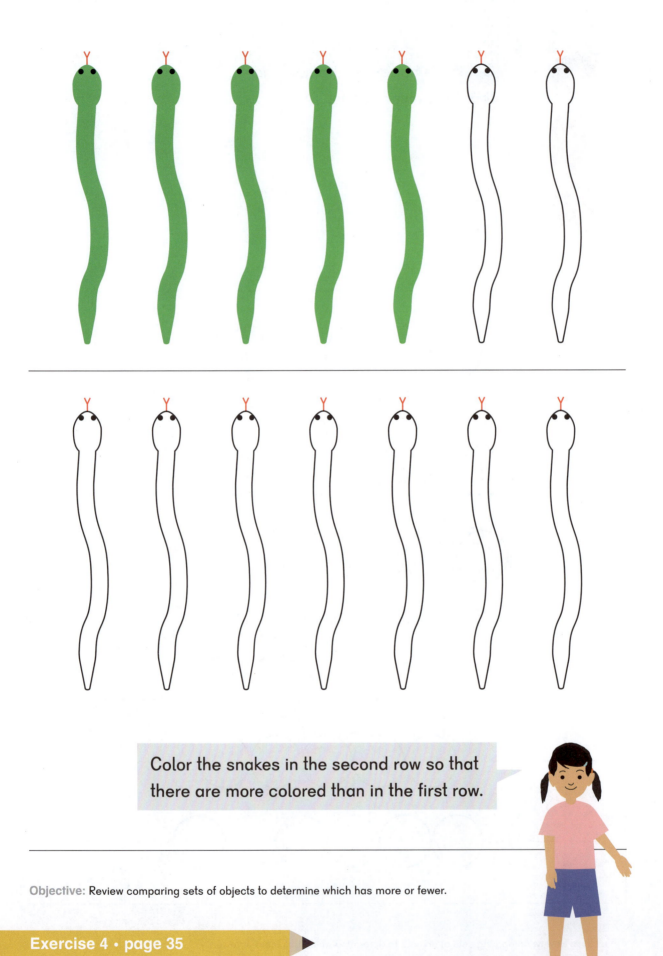

Color the snakes in the second row so that there are more colored than in the first row.

Objective: Review comparing sets of objects to determine which has more or fewer.

Exercise 4 • page 35

Lesson 5
Practice

P 5

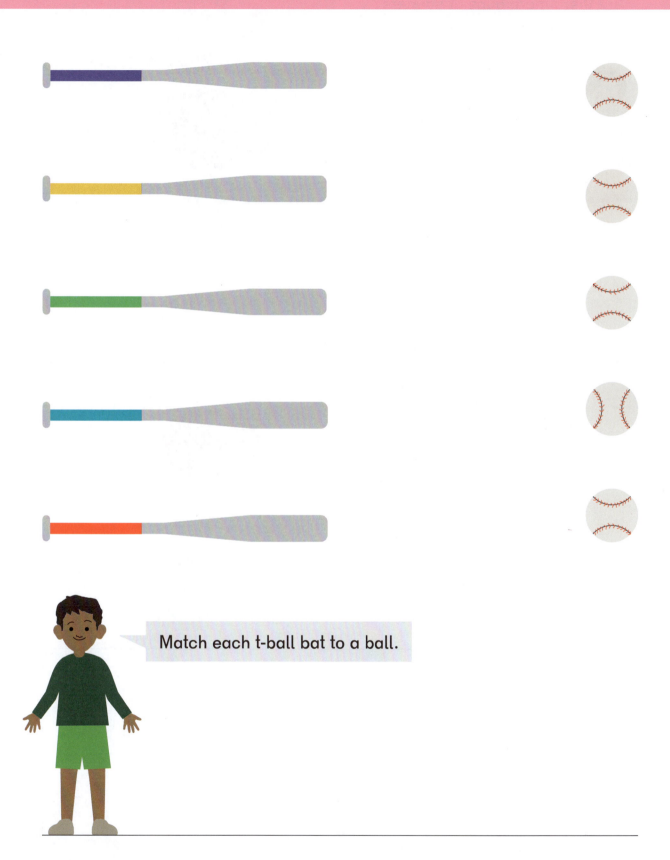

Match each t-ball bat to a ball.

Objective: Practice.

Are there more cups or saucers?
Circle the answer below.

There are more 🟡 .   There are more 🥛 .

Objective: Practice.

Are there fewer cats or balls of yarn? Circle the answer below.

There are fewer 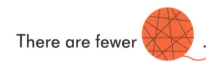.   There are fewer .

Objective: Practice.

Color the second tower so that it has fewer cubes colored than the first tower.

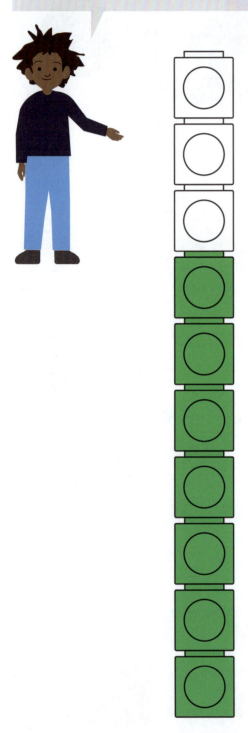
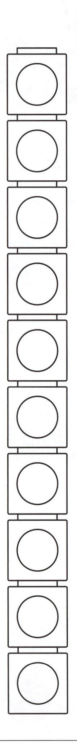

Objective: Practice.

Exercise 5 • page 37

Chapter 11

Compose and Decompose

Look and talk.
3 gray puppies and 2 brown puppies make 5 puppies altogether.

Lesson 1
Altogether — Part 1

Look and talk.
2 bears on a shelf and 2 bears on the floor makes 4 bears in all.

Objective: Use pictures to compose numbers to five.

Exercise 1 • page 39

Lesson 2
Altogether — Part 2

Circle the hands that show 4 fingers altogether.

Objective: Use fingers to compose numbers to five.

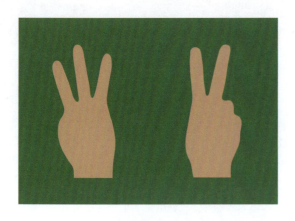

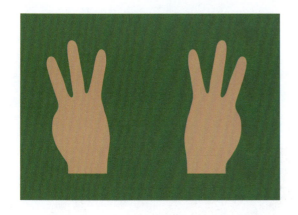

Circle the hands that show 5 fingers altogether.

Objective: Use fingers to compose numbers to five.

Circle the ladybugs that show 5 dots.

Objective: Use pictures to compose numbers to five.

Exercise 2 • page 41

Lesson 3
Show Me

Color in the boxes with 2 different colors to show the 2 sets.
The first one has been done for you.

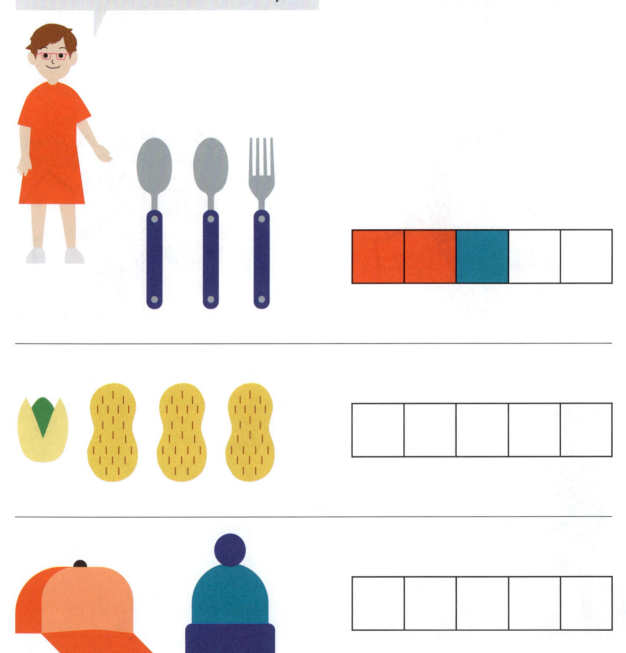

Objective: Record composition of numbers to five using five-frames.

Objective: Record composition of numbers to five using five-frames.

Color in the boxes with 2 colors to show the 2 parts. Then circle how many in all.

4   3   2

2   5   3

5   4   3

Objective: Show parts of five on five-frames. Compose numbers within five.

Exercise 3 • page 43

Lesson 4
What's the Other Part? — Part 1

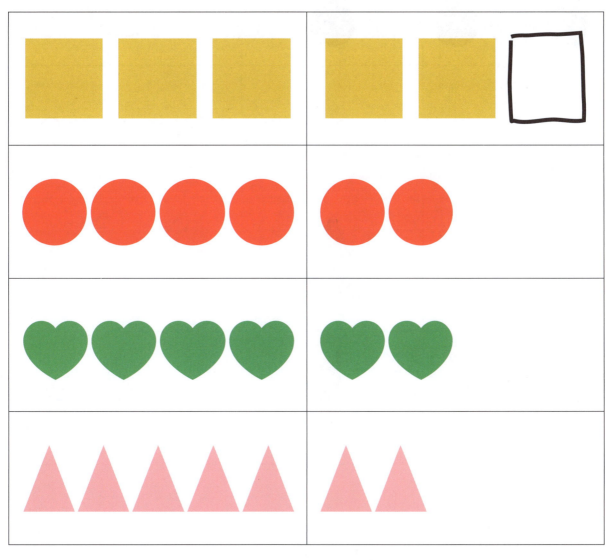

Draw the missing part of each set. The first one is done for you.

Objective: Find how many in the part not given.

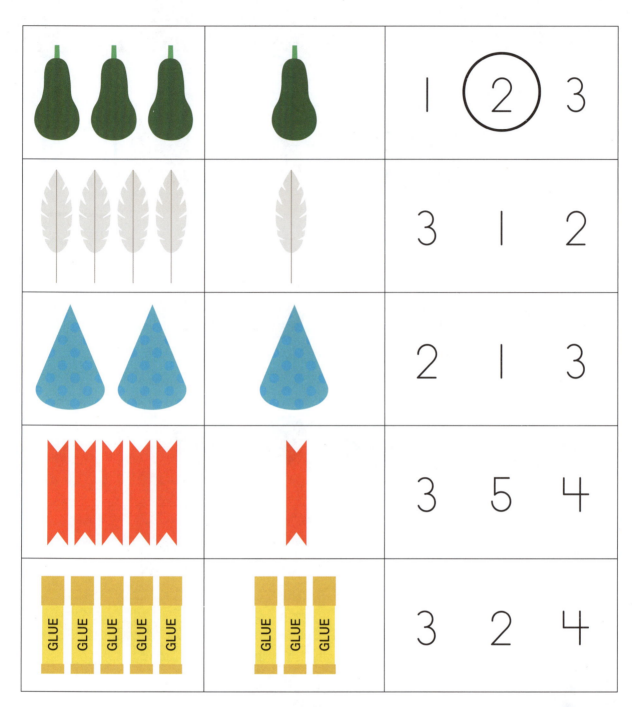

How many are missing in the second set? The first one is done for you.

Objective: Find how many in the part not given.

Exercise 4 • page 45

Lesson 5
What's the Other Part? — Part 2

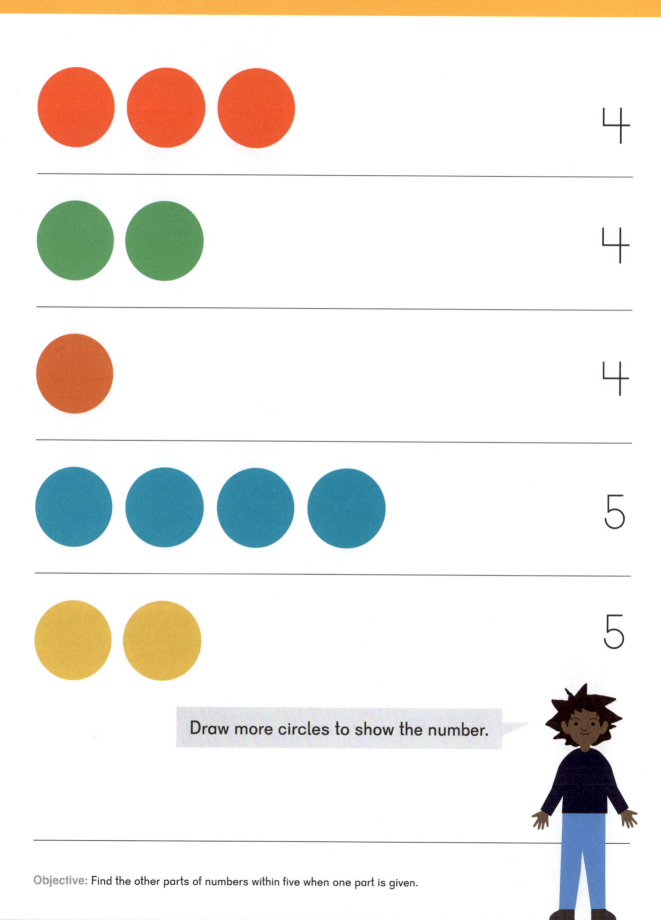

Draw more circles to show the number.

Objective: Find the other parts of numbers within five when one part is given.

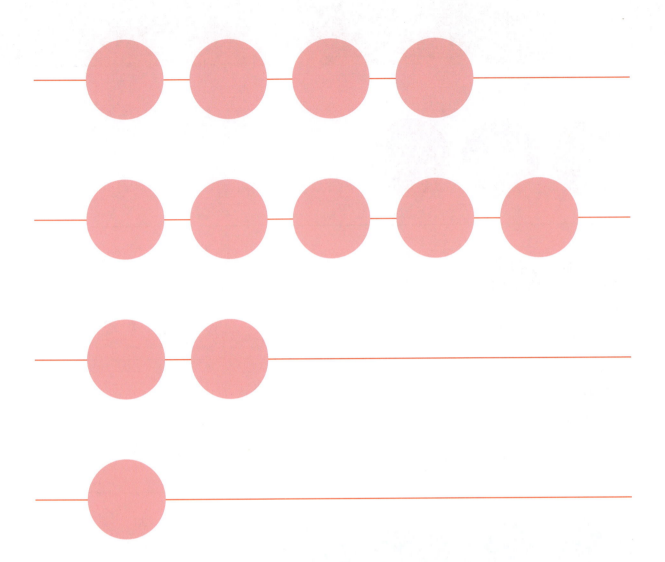

Draw beads so that there are 5 on each necklace.

Exercise 5 • page 49

Lesson 6
Practice

Circle the hands that show 5 fingers altogether.

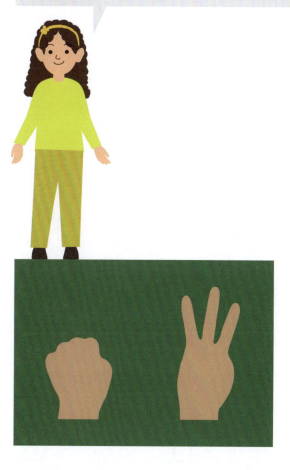
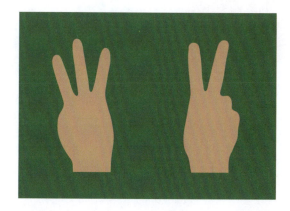

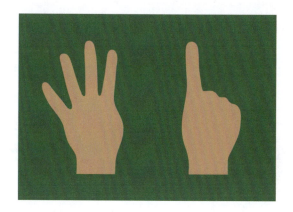

Objective: Practice.

Circle the number of balloons altogether.

   4   3   2

 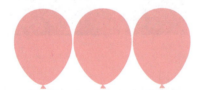   4   2   5

    5   4   3

Objective: Practice.

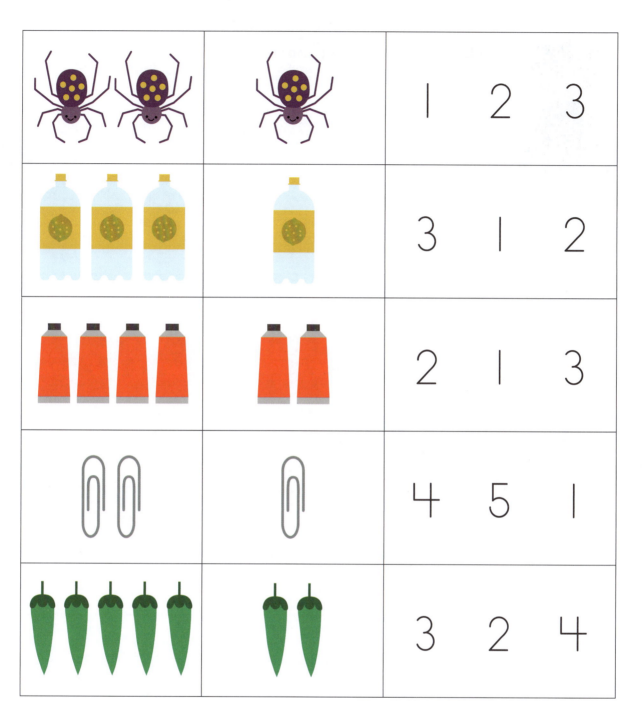

How many are missing in the second set?

Objective: Practice.

Draw more beads so there are 5 beads on each string.

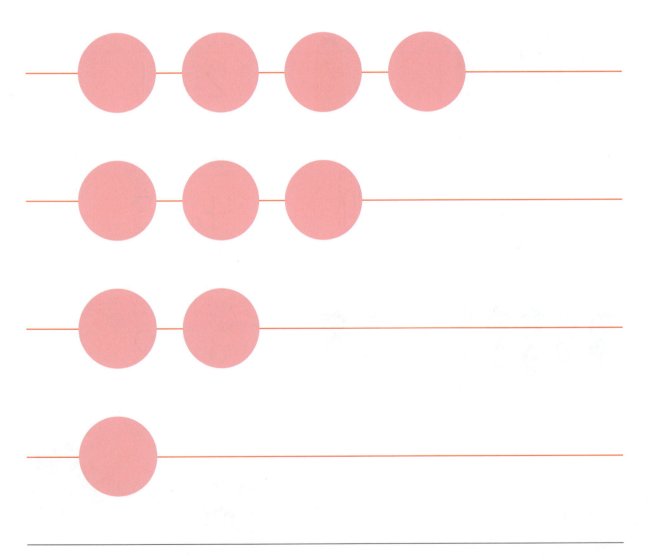

Objective: Practice.

Exercise 6 • page 53

Chapter 12

Explore Addition and Subtraction

Lesson 1
Add to 5 — Part 1

Match each picture showing 2 parts to the picture that shows the whole. The first one is done for you.

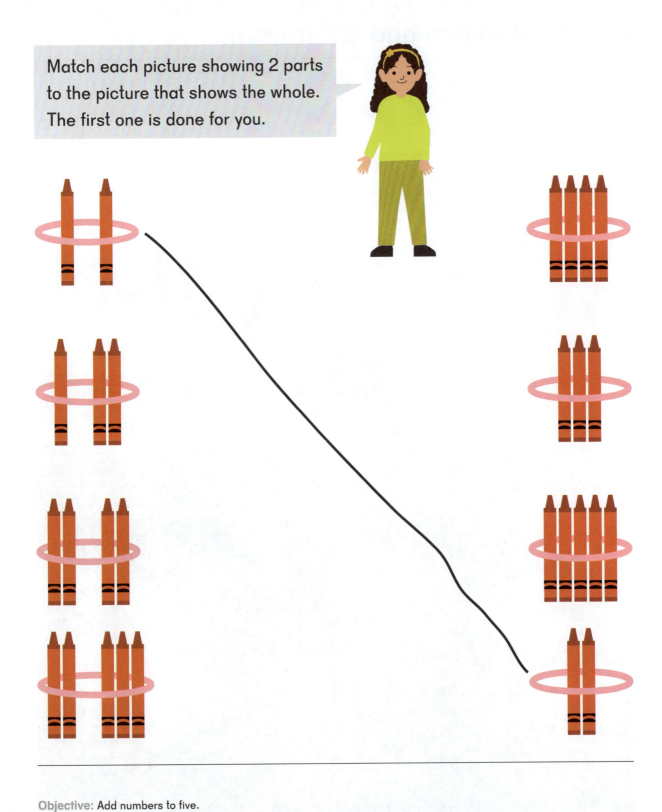

Objective: Add numbers to five.

Exercise 1 • page 55

Lesson 2
Add to 5 — Part 2

Look and talk.
1 shoe off and 1 shoe on makes 2 shoes in all.

Diddle, diddle, dumpling, my son John,
Went to bed with his trousers on.
One shoe off, and one shoe on,
Diddle, diddle, dumpling, my son John.

Objective: Add numbers to five.

    2   3   4

    2   3   4

    3   4   5

Circle the number that shows how many altogether.

Objective: Add numbers to five.

Exercise 2 • page 57

Lesson 3
Two Parts Make a Whole

3

Make up a story for each picture.

Objective: Create addition stories from pictures.

Make up a story for each picture, then circle how many in all.

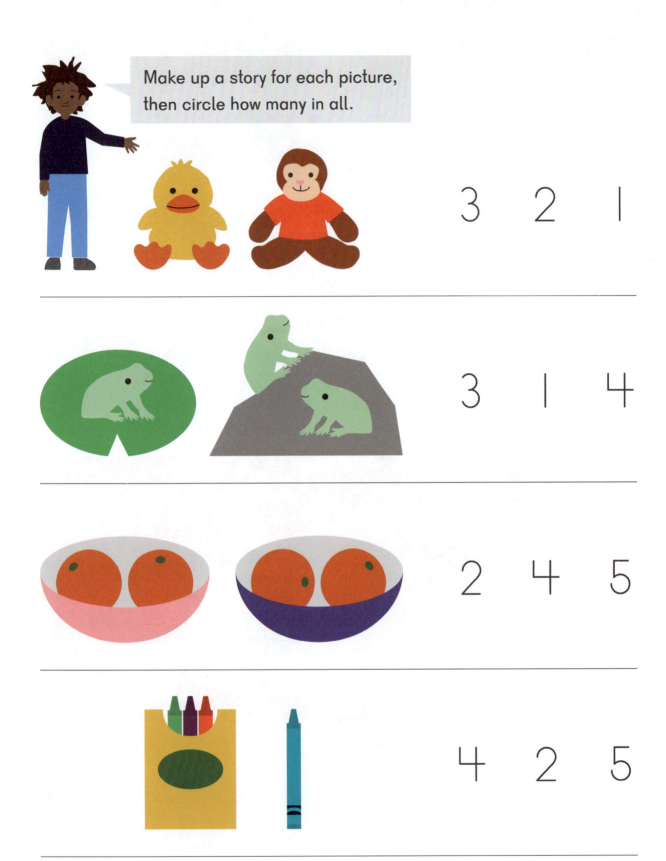

3 2 1

3 1 4

2 4 5

4 2 5

Objective: Create addition stories from pictures.

Exercise 3 • page 61

Lesson 4
How Many in All?

Little Red Riding Hood picked 1 pink flower.
She then picked 2 yellow flowers.
How many flowers did she pick altogether?

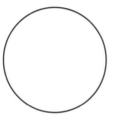 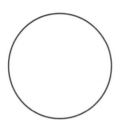 | 3   4   5

Little Red Riding Hood saw 2 tall trees and 2 short trees before she started walking in the forest.
How many trees did she see before she got to the forest?

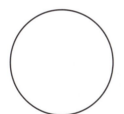 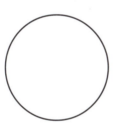 | 3   4   5

Draw dots in the circles to go with the addition stories, then circle the correct number for the whole.

Objective: Add numbers to five.

Exercise 4 • page 63

Lesson 5
Subtract Within 5 — Part 1

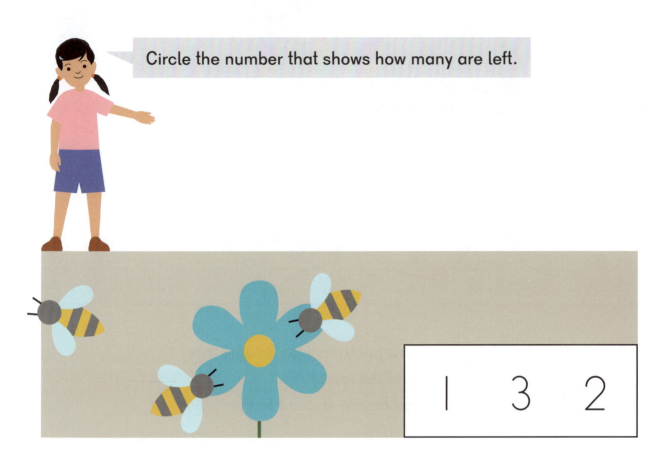

Objective: Subtract numbers within five.

Circle the number that shows how many are left.

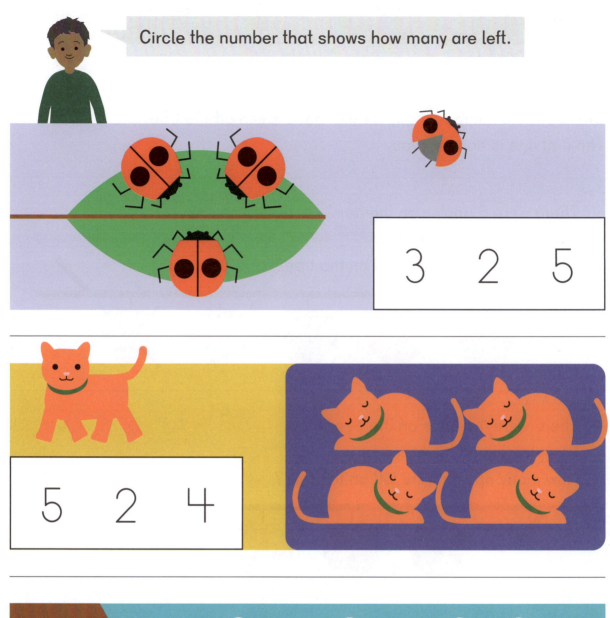

Objective: Use objects and pictures to subtract numbers within five.

Exercise 5 • page 65

Lesson 6
Subtract Within 5 — Part 2

Cross out the number of apples that have fallen off the tree.
The first one is done for you.

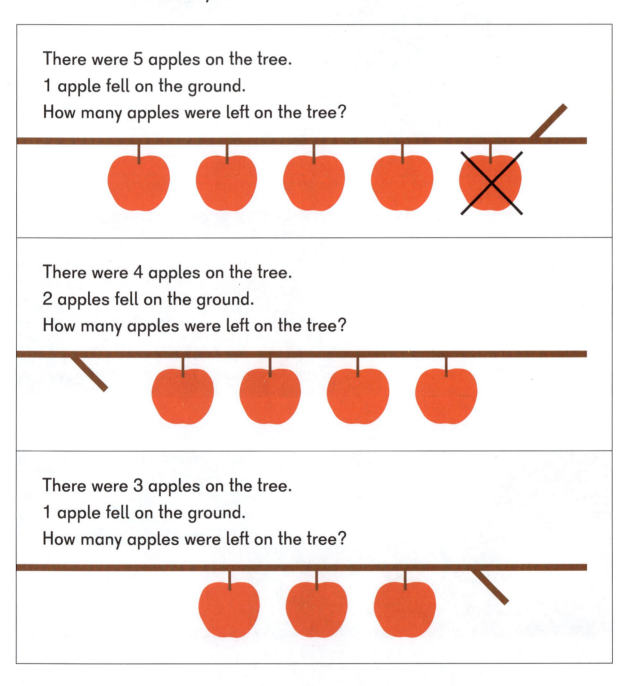

There were 5 apples on the tree.
1 apple fell on the ground.
How many apples were left on the tree?

There were 4 apples on the tree.
2 apples fell on the ground.
How many apples were left on the tree?

There were 3 apples on the tree.
1 apple fell on the ground.
How many apples were left on the tree?

Objective: Subtract numbers within five.

Exercise 6 • page 67

Lesson 7
How Many Are Left?

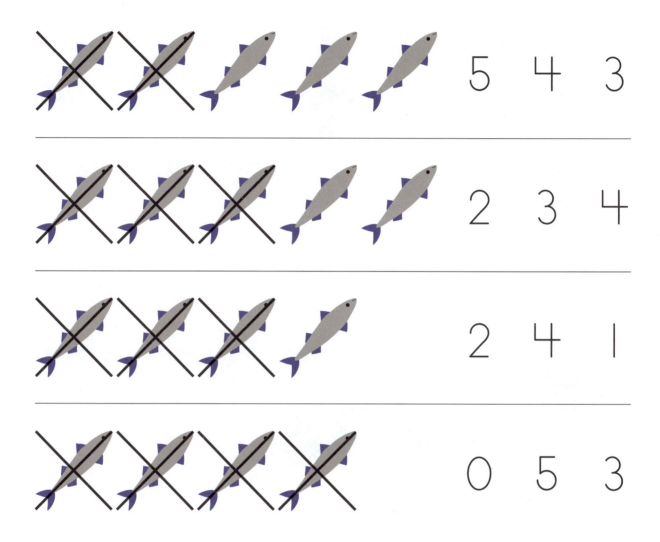

Circle the number that shows how many are left. Make up a story for each set of pictures.

Objective: Subtract numbers within five.

Exercise 7 • page 69

Lesson 8
Practice

P 8

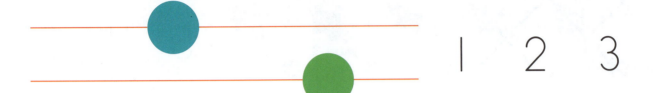

1 2 3

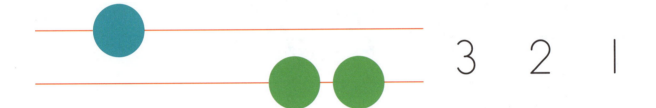

3 2 1

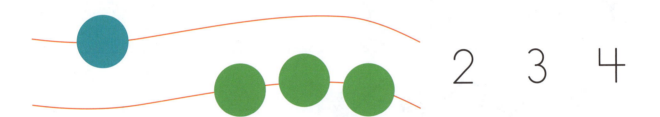

2 3 4

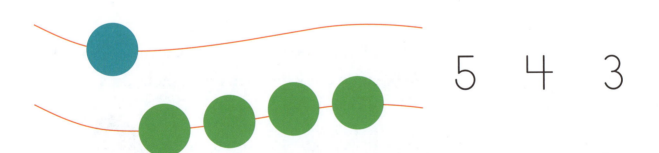

5 4 3

Circle the number of beads in all.

Objective: Practice.

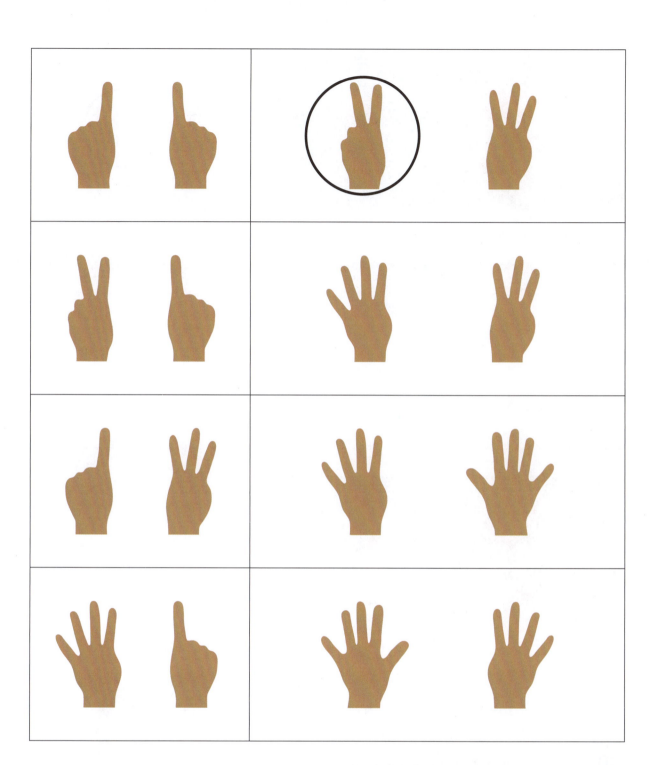

Circle the hand that shows the correct number of fingers in all. The first one is done for you.

Objective: Practice.

    3 ④ 5

    3   5   2

    4   5   3

Cross out crabs that are going away, then circle the number that shows how many are left. The first one has been done for you.

Objective: Practice.

Exercise 8 • page 71

Chapter 13

Cumulative Review

Review 1
Match and Color

R 1

Cross out the one in each row that is not the same.

Objective: Identify objects that are the same and not the same.

Objective: Draw squares and circles. Recognize the colors blue, red, brown, green, orange, and yellow.

Color the picture.

Objective: Recognize the colors blue, red, brown, green, orange, yellow, and pink.

Exercise 1 • page 73

Review 1 Match and Color

Review 2
Big and Small

Objective: Compare by size and color.

Exercise 2 • page 75

Review 3
Heavy and Light

R 3

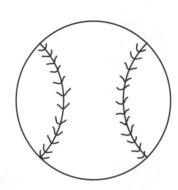

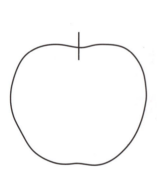
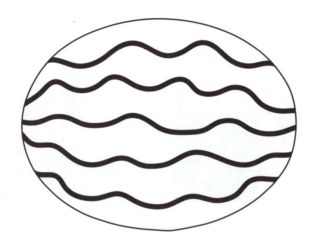

Color the heaviest one blue and circle the lightest one in each row.

Objective: Compare by weight.

Exercise 3 • page 77

Review 4
Count to 5

Count and circle the set of 3 squares.

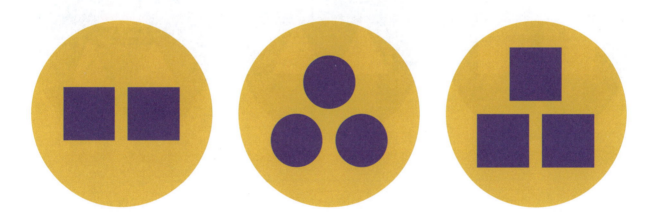

Objective: Count sets of objects up to five with one-to-one correspondence and identify squares.

Count and circle the set of 2 triangles.

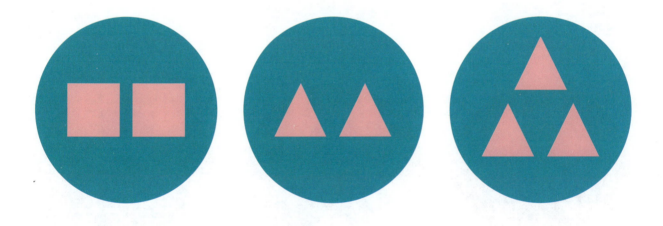

Objective: Count sets of objects up to five with one-to-one correspondence and identify triangles.

Exercise 4 • page 79

Review 5
Count 5 Objects

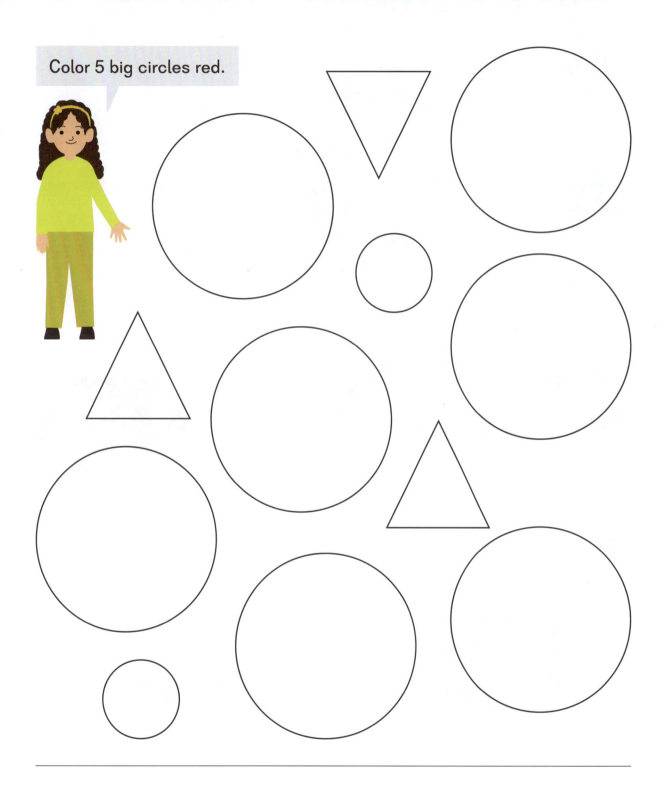

Objective: Count sets of objects up to five with one-to-one correspondence and identify circles.

Exercise 5 • page 81

Review 6
0

Spot has 0 doggie biscuits.
Draw a line to match Spot to his bowl.

Objective: Understand that an empty set is a set of zero.

Exercise 6 • page 83

Review 7
Count Beads

R 7

Match the number of beads on a necklace to the number.

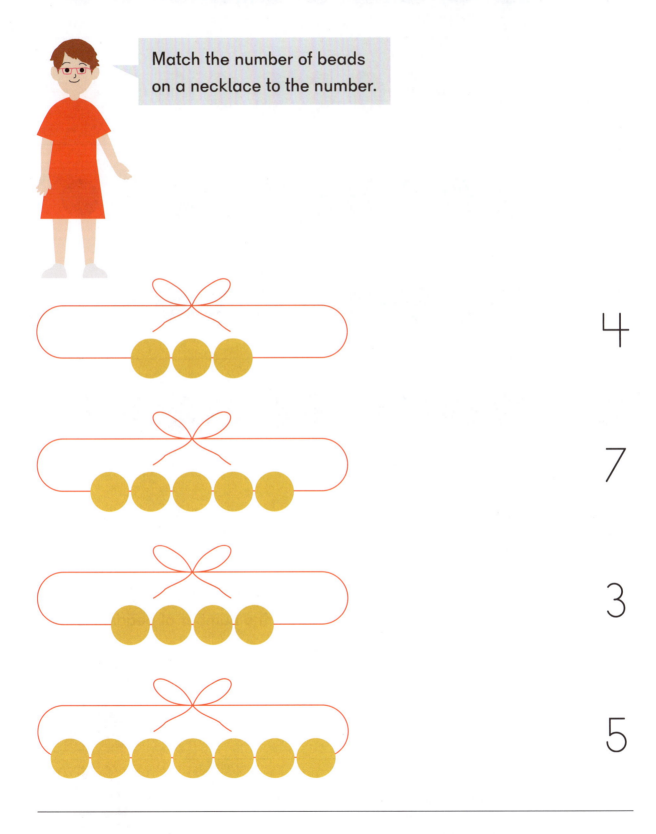

Objective: Count with one-to-one correspondence. Recognize numerals 0 to 10.

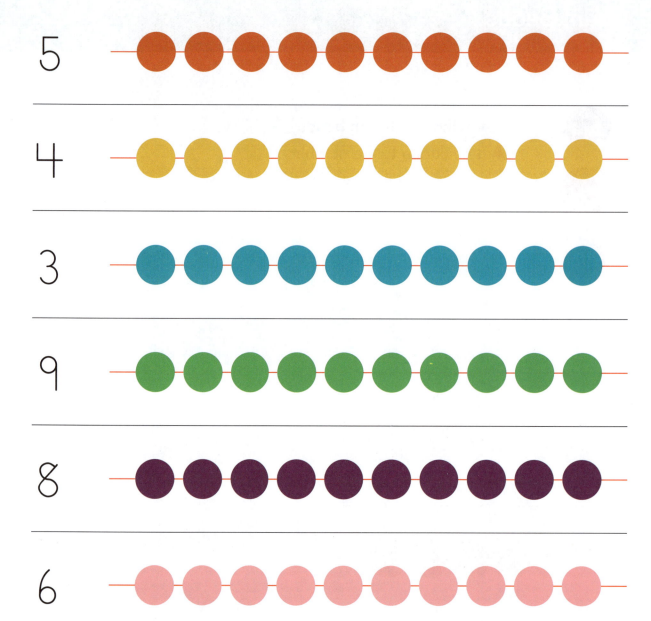

Circle the number of beads shown.

Objective: Count with one-to-one correspondence. Recognize numerals 0 to 10.

Exercise 7 • page 85

Review 8
Patterns

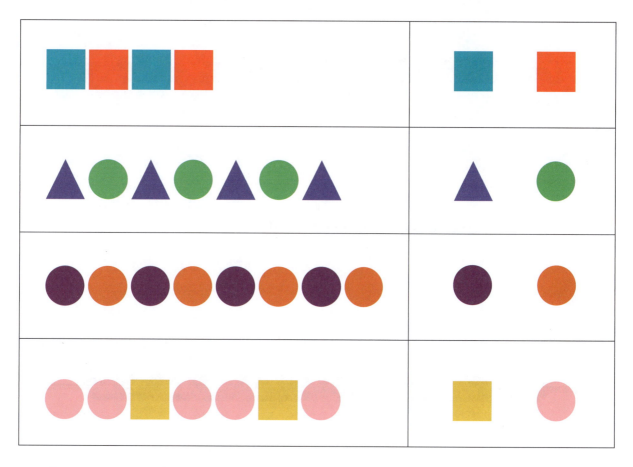

What comes next? Circle it.

Objective: Identify and extend patterns.

Exercise 8 • page 87

Review 9
Length

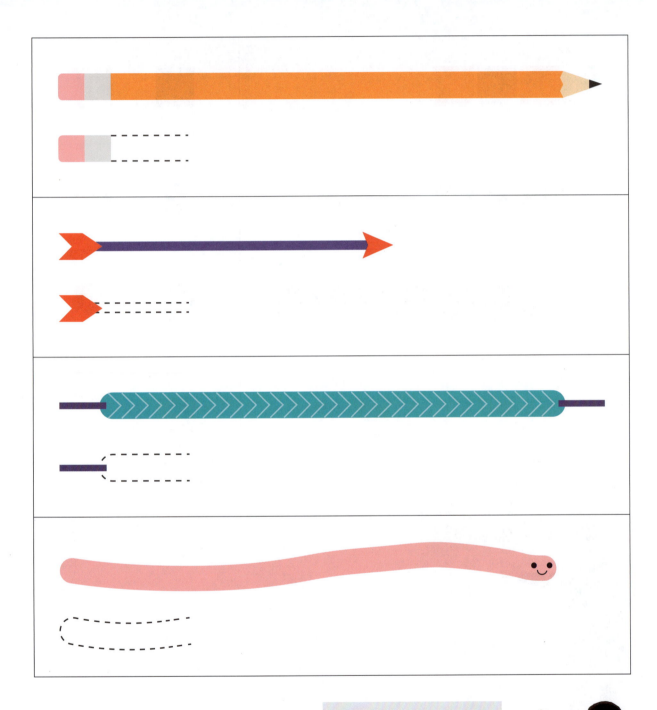

Draw a shorter one.

Objective: Compare objects by length.

Exercise 9 • page 89

Review 10
How Many?

Circle the number that shows how many.

   2   3   4

   4   5   6

   4   3   2

   3   4   6

   2   4   6

Objective: Subitize dice patterns.

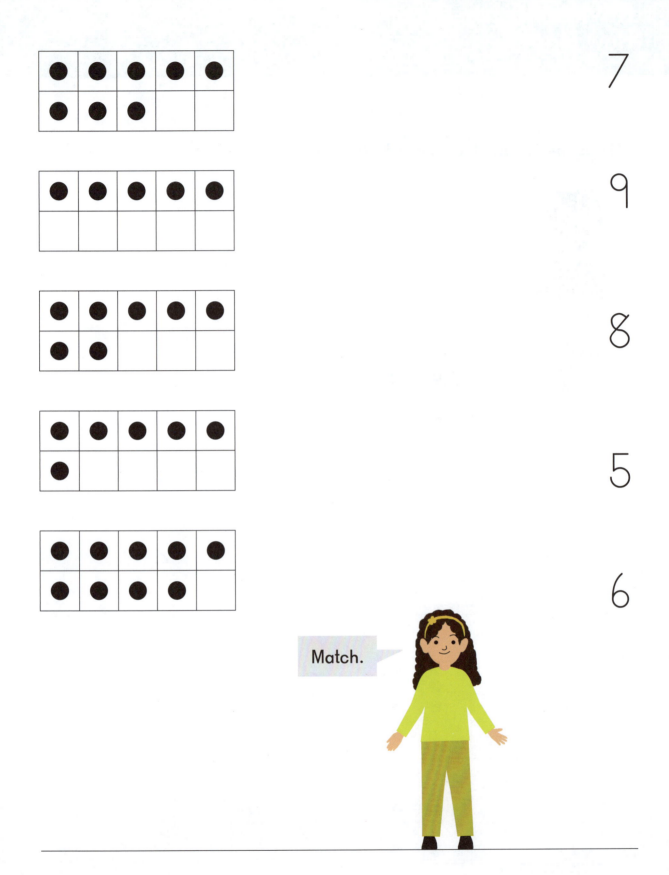

Review 11
Ordinal Numbers

The first triangle is purple.
Circle the fourth triangle.

Objective: Identify first through fifth from a starting point.

Exercise 11 • page 93

Review 12
Solids and Shapes

R (12)

Circle the cubes.
Cross out the cylinders.

Objective: Recognize spheres, cubes, and cylinders.

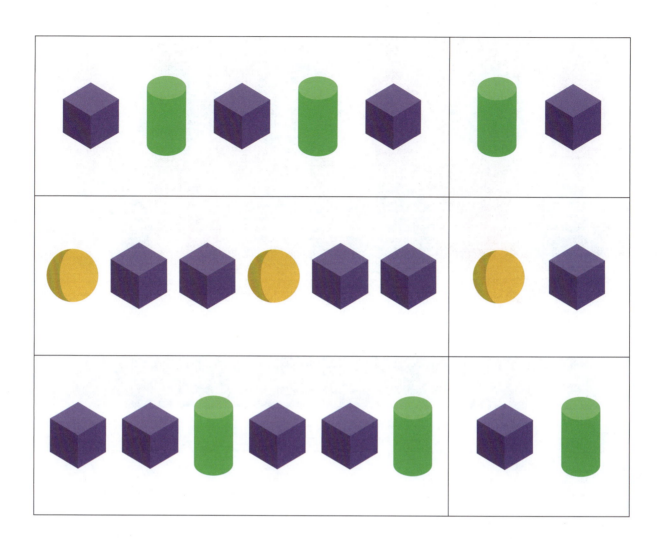

What comes next? Circle it.

Objective: Extend patterns.

Review 12 Solids and Shapes

Solids		

Color in the graph to show how many.

Objective: Make a graph from a picture.

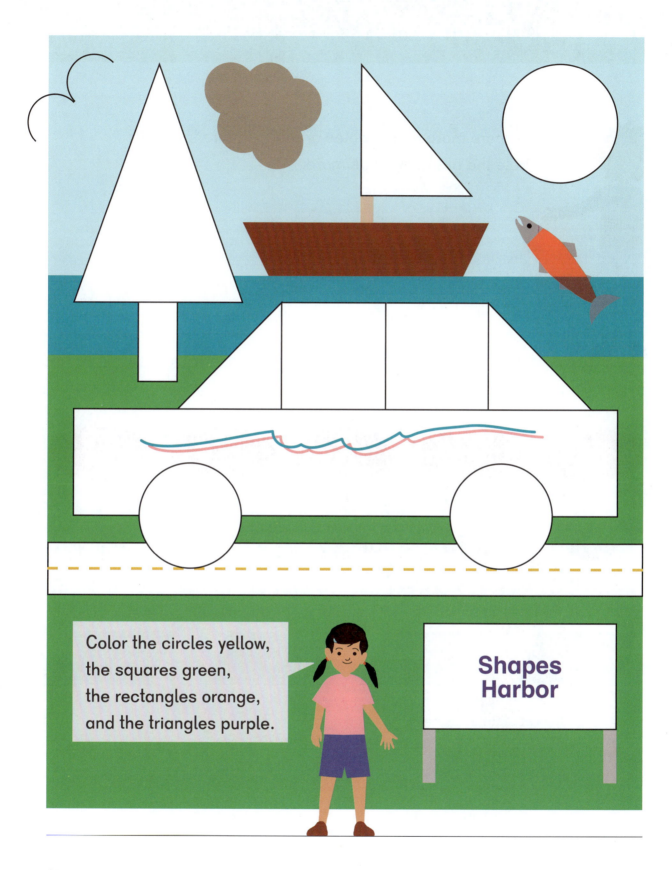

Color the circles yellow, the squares green, the rectangles orange, and the triangles purple.

Shapes Harbor

Objective: Recognize circles, squares, rectangles, and triangles.

Exercise 12 • page 95

Review 13
Which Set Has More?

Are there more round tables or more square tables? Circle the group that has more.

Objective: Compare two sets of objects to find which has more.

Exercise 13 • page 97

Review 14
Which Set Has Fewer?

Are there fewer long pencils or fewer short pencils? Circle the group that has fewer.

Objective: Compare two sets of objects to find which has fewer.

Exercise 14 • page 99

Review 15
Put Together

Match.

Objective: Identify the whole when two parts are given.

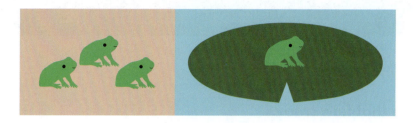 1 3 4

 5 3 4

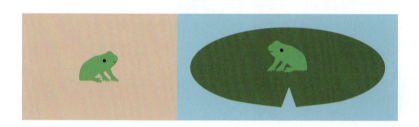 2 3 4

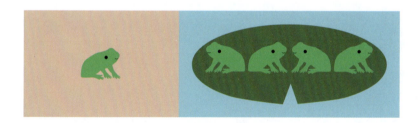 3 4 5

Circle the number that shows how many altogether.

Objective: Identify the whole when two parts are given.

Exercise 15 • page 101

Review 16
Subtraction

Cross out the number of bats that flew away.
Circle the number left.

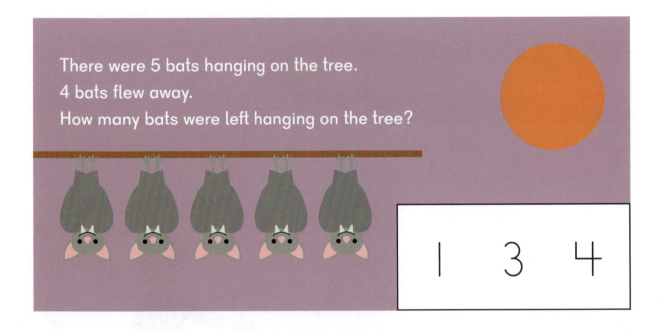

Objective: Subtract numbers within five.

Exercise 16 • page 103

Looking Ahead 1
Sequencing — Part 1

Circle the picture that comes first.

Objective: Use logic to determine the sequence of events.

Looking Ahead 2
Sequencing — Part 2

Circle the picture that might come next.

Objective: Use logic to determine the sequence of events.

Circle the picture that might come next.

Looking Ahead 3
Categorizing

Find 4 things that go together and tell why.

Objective: Categorizing.

Looking Ahead 4
Addition

Circle the number.
The first one is done for you.

2 and 1 make | 2 ③ 1 |

2 + 1 = | 2 ③ 1 |

3 and 2 make | 5 4 3 |

3 + 2 = | 5 4 3 |

Objective: Add two parts to find the whole.

Circle the correct number.

1 and 1 make | 3 2 4 |

1 + 1 = | 3 2 4 |

2 and 2 make | 4 5 3 |

2 + 2 = | 4 5 3 |

Objective: Add two parts to find the whole.

Looking Ahead 5
Subtraction

Circle the number.
The first one is done for you.

2   take away   1   is   3 ① 4

2   −   1   =   3 ① 4

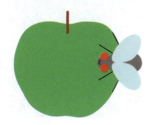

3   take away   2   is   1   3   5

3   −   2   =   1   3   5

Objective: Take away one part from the whole to find the other part.

Circle the number.

4   take away   2   is | 2 4 1 |

4   −   2   = | 2 4 1 |

5   take away   2   is | 3 4 2 |

5   −   2   = | 3 4 2 |

Objective: Take away one part from the whole to find the other part.

Looking Ahead 6
Getting Ready to Write Numerals

Trace the shapes.
Then color them.

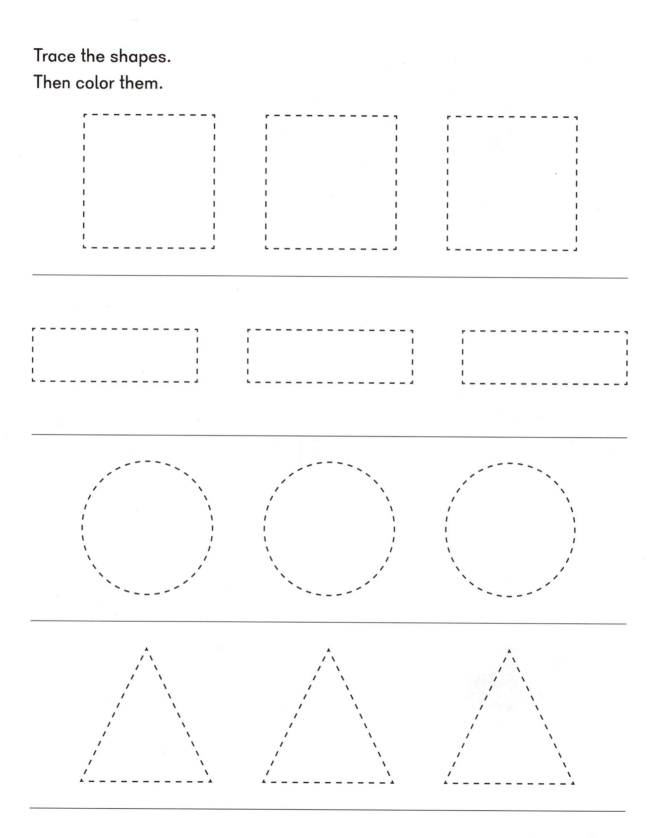

Objective: Drawing vertical, horizontal, curved, and diagonal lines.

Looking Ahead 7
Reading and Math

Circle the number of letters in the word.

 dog   1   2   3

 frog   4   3   5

 truck   2   5   4

Objective: Find the number of letters in simple words.